I Am Earth

HOUSE OF LORE

Earth

Rebecca and James McDonald

I Am Earth
An Earth Day Book for Kids

ISBN: 978-0-9982949-0-2

First House of Lore paperback edition, 2016

Visit us at www.HouseOfLore.net

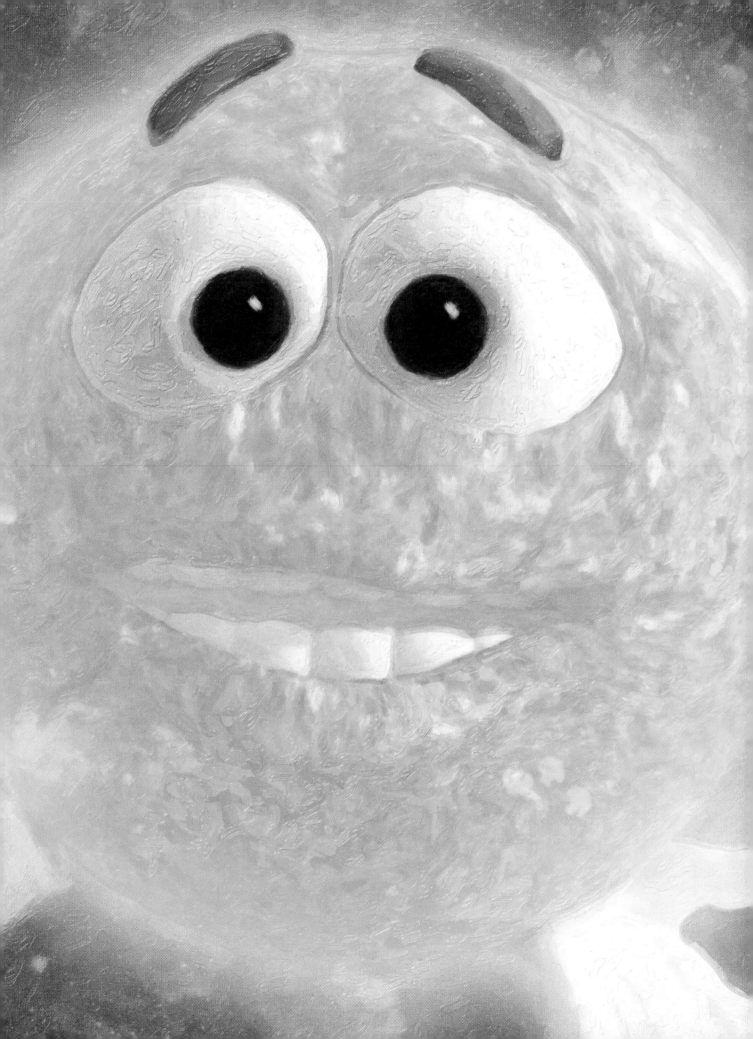

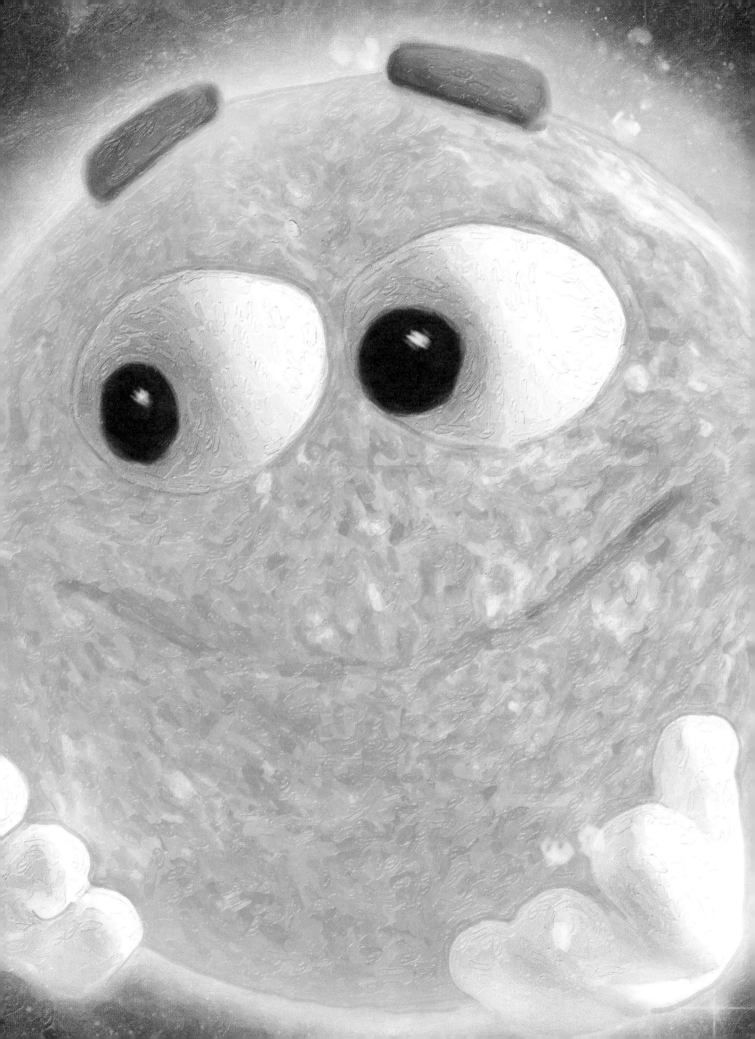

And the white is from
the clouds and snow.

Clouds drop water on the land below. And when it gets cold, the clouds drop snowflakes, dusting everything white.

Rain and snow fill up the lakes and rivers, giving all living things water to drink. Even the food you eat is grown in my rich brown soil.

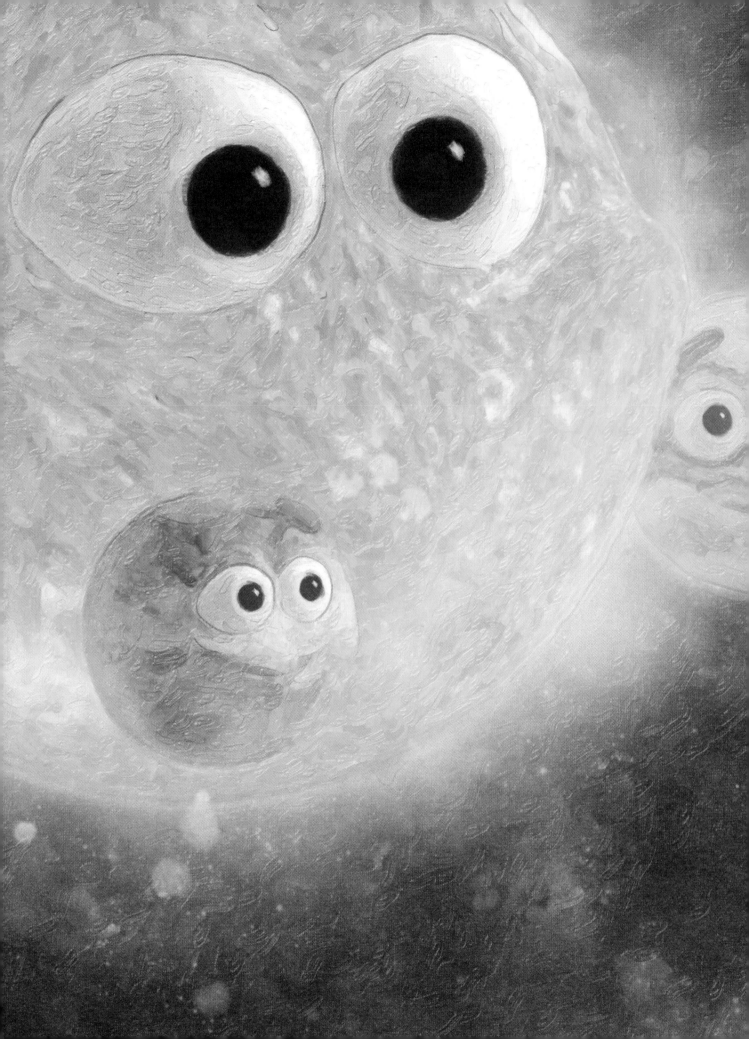

Here are some things you can do to help the Earth be the best it can be!

Recycle- Find out what can be recycled and start recycling.

Don't Litter- Put garbage in garbage cans.

Reuse- Use things more than once if you can, like paper bags and glass jars.

Save energy- Turn off the lights when you don't need them.

Don't waste water- Even though there's lots of water on the Earth, there's only a small amount that we can actually drink, so let's not waste it!

Can you think of other ways to help the Earth stay a happy, healthy home?

And remember to tell all of your family and friends, so they can help too!

CHECK OUT THESE OTHER TITLES FROM HOUSE OF LORE